Charlie explores

whole number

7

Sarah Reeves

Sarah's Books Pty Ltd, PO Box 1904 Armidale NSW 2350

First Edition, 2024
First published, 2024

ISBN: 978-1-7635799-2-7
Independently published.

Printed in Sydney, NSW, Australia, if purchased in Australia.
eBook also available.

A catalogue record for this book is available from the National Library of Australia.

This book was created, written and illustrated in Armidale NSW, which is Anaiwan country.

Links to the Australian Mathematics Curricula

Australian Curriculum Mathematics F - 10
Foundation - AC9MFN06 - represent practical situations involving equal sharing and grouping with physical and virtual materials and use counting or subitising strategies.

NSW Mathematics K-10 Curriculum
Early Stage 1 - MAE-RWN-01 - demonstrates an understanding of how whole numbers indicate quantity.

Victorian Mathematics F - 6 Curriculum
Foundation - VC2MFN06 - represent practical situations that involve equal sharing and grouping with physical and virtual materials and use counting or subitising strategies.

Links to curricula:
The Australian Curriculum, Assessment and Reporting Authority (ACARA) Mathematics F-10 Version 9.0
The NSW Education Standards Authority (NESA) Mathematics K-10 Curriculum (February 2024)
The Victorian Curriculum and Assessment Authority Mathematics Foundation to Level 6 V 2.0

This book is dedicated to Katie.

Your creativity is an inspiration,

Thank you for always being excitied & ready to listen.

Charlie plays in the sandpit

and writes the word

Seven.

Charlie writes the numeral

7

Charlie finds 7 leaves.

Charlie then finds 7 rocks.

And 7 sticks.

Charlie arranges 7 shells.

Charlie loves to count to 7.

Charlie arranges

4 rocks and 3 shells.

Charlie still has 7.

Charlie then uses

3 leaves, 2 rocks

and 2 sticks.

Charlie still has 7.

Next Charlie has

1 leaf, 2 rocks,

2 sticks & 2 shells

to make 7.

Numbers can be expressed

in many different ways

and with lots of different

objects and things.

The End.

Other books available in the Charlie Maths series:

Charlie counts to five, on a picnic
ACARA Mathematics Curriculum link AC9MFN02
NSW Mathematics K-10 Curriculum link MAE-RWN-001 &
MAE-RWN-002
Victorian Mathematics F - 6 Curriculum link VC2MFN02

Charlie loves to share, exploring odd and even numbers
ACARA Mathematics Curriculum link AC9MFN06
NSW Mathematics K-10 Curriculum link MAE-FG-02 & MA1-FG01
Victorian Mathematics F - 6 Curriculum link VC2MFN06 & VC2M3N01

Charlie looks into greater than, less than and equal to
ACARA Mathematics Curriculum link AC9MFN06
NSW Mathematics K-10 Curriculum link MAE-RWN-02
Victorian Mathematics F - 6 Curriculum link VC2MFN01

Charlie finds a pattern, in the sandpit
ACARA Mathematics Curriculum link AC9MFA01
NSW Mathematics K-10 Curriculum link MAE-FG-01
Victorian Mathematics F - 6 Curriculum link VC2MFA01

Charlie goes for a walk, introducing algebra, with oranges
ACARA Mathematics Curriculum link AC9MFA01 & AC9M1A02
NSW Mathematics K-10 Curriculum link MAE-FG-02
Victorian Mathematics F - 6 Curriculum link VC2MFA01 & VC2MFA02

Other books available in the Charlie Maths series:

Charlie loves to cook, exploring measurement
ACARA Mathematics Curriculum link AC9MFM01
NSW Mathematics K-10 Curriculum link MAE-CSQ01
Victorian Mathematics F - 6 Curriculum link VC2MFM01

Charlie explores fractions
ACARA Mathematics Curriculum link AC9M2M02
NSW Mathematics K-10 Curriculum link MAE-GM-03
Victorian Mathematics F - 6 Curriculum link VC2M2M02

Charlie finds Geometry, in nature
ACARA Mathematics Curriculum link AC9MFSP01
NSW Mathematics K-10 Curriculum link MAE-2DS-01 & MA1-2DS-01
Victorian Mathematics F - 6 Curriculum link VC2MFSP01

Links to curricula:
The Australian Curriculum, Assessment and Reporting Authority
(ACARA) Mathematics F-10 Version 9.0
The NSW Education Standards Authority (NESA) Mathematics K-10
Curriculum (February 2024)
The Victorian Curriculum and Assessment Authority Mathematics
Foundation to Level 6 V 2.0

www.ingramcontent.com/pod-product-compliance
Lightning Source LLC
Chambersburg PA
CBHW042120060426
42449CB00030B/38